A LITTLE BIT OF DINOSAUR!

by Elleen Hutcheson and Darcy Pattison
illustrated by John Joven

A Little Bit of Dinosaur
by Elleen Hutcheson and Darcy Pattison
illustrated by John Joven

Text © 2021 Elleen Hutcheson and Darcy Pattison
Illustrations © 2021 Mims House, LLC

Mims House
1309 Broadway
Little Rock, AR 72202
USA

MimsHouseBooks.com

Publisher's Cataloging-in-Publication Data

Names: Hutcheson, Elleen, author. | Pattison, Darcy, author. | Joven, John, illustrator.
Title: A little bit of dinosaur! / By Elleen Hutcheson and Darcy Pattison; illustrated by John Joven.
Description: Little Rock, AR: Mims House, 2020. | Summary: This humorous story follows a calcium atom as it journeys from dry bones to your jawbone.
Identifiers: LCCN 2020913292 | ISBN 978-1-62944-153-5 (Hardcover) | 978-1-62944-154-2 (pbk.) | 978-1-62944-155-9 (ebook) | 978-1-62944-156-6 (audio)
Subjects: LCSH Life cycles (Biology)--Juvenile literature. | Calcium in the body--Juvenile literature. | Teeth--Juvenile literature. | Bones--Juvenile literature. | Dinosaurs--Juvenile literature. | Humorous stories. | CYAC Life cycles (Biology). | Calcium in the body. | Teeth. | Bones. | Dinosaurs. | BISAC JUVENILE NONFICTION / Science & Nature / Biology | JUVENILE NONFICTION / Science & Nature / Anatomy & Physiology | JUVENILE NONFICTION / Science & Nature / Fossils | JUVENILE NONFICTION / Animals / Dinosaurs & Prehistoric Creatures
Classification: LCC QH501.H88--dc23 | DDC 372.3/57--dc23

You have
a little bit of
Tyrannosaurus
rex
in your
jawbone!

Don't believe me? It's all your mother's fault.

Listen up.
Here's how
it happened.

Once, in days of old, a Tyrannosaurus rex roamed the lands of North America. When his days on earth were done, he died.

Layer upon layer of mud buried his bones.

Time ticked by. The mud hardened into rock. The land lifted to make the Rocky Mountains. Centuries slid by. Rain fell, eroding the rock, washing it away, bit by bit.

Slowly, slowly, a T. rex toe bone was uncovered. Rain fell, eroding the bone, washing away a little bit of dinosaur— some calcium. Water swept that little bit of calcium down the valley to the Arkansas River.

COLORADO

KANSAS

OKLAHOMA

The river carried the little bit of dinosaur across Colorado, Kansas, and Oklahoma, and might have carried it to the Mississippi River, except—it got sucked up by a pump that irrigated a cornfield.

The little bit of calcium that used to be part of a T. rex was absorbed by a corn plant and became part of an ear of corn.

**The farmer harvested the corn,
and fed it to his cow.**

The cow's body used
the little bit of
dinosaur
to make milk.

The farmer milked the cow
and sent the milk to a dairy.

The dairy used
the milk to make
cheese.

A trucker drove the cheese to the store. Your mother bought the cheese and made you a sandwich using cheese that had calcium that used to be in a dinosaur.

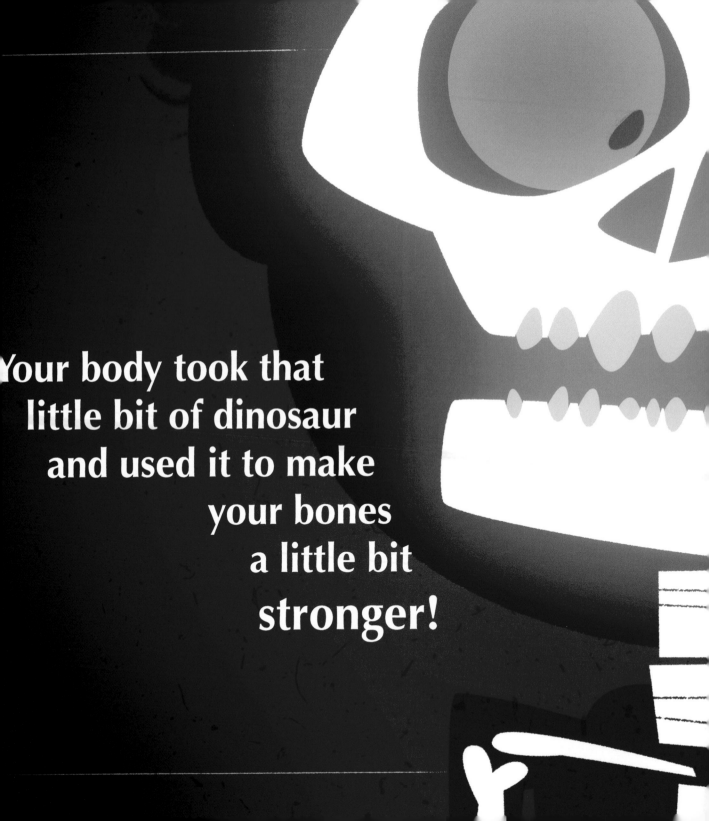

Your body took that little bit of dinosaur and used it to make your bones a little bit **stronger!**

You have
a little bit of
Tyrannosaurus rex
in your jawbone.

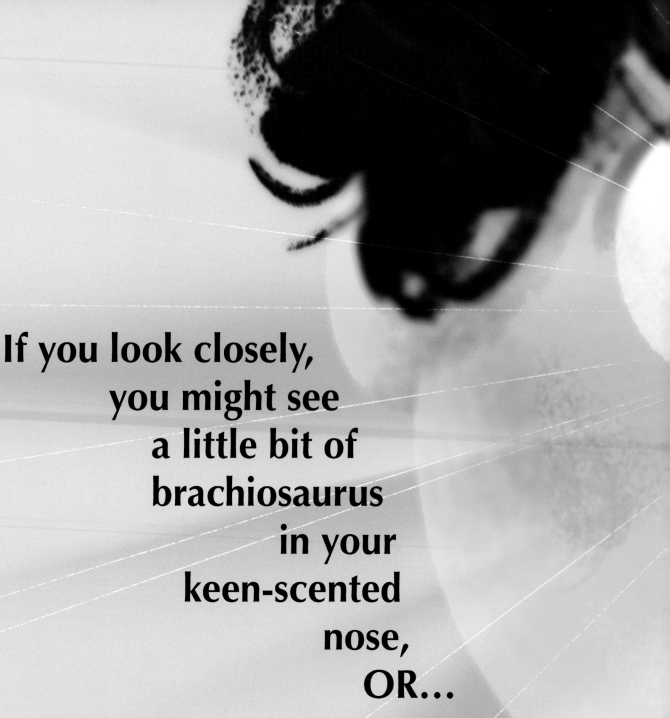

**If you look closely,
you might see
a little bit of
brachiosaurus
in your
keen-scented
nose,
OR...**

...a little bit of stegosaurus in your strong spine, OR...

...a little bit of velociraptor in your grasping hands, OR...

...a little bit of apatosaurus in your big toe. And when...

...your days on earth are done, and your body returns to the land, that little bit of dinosaur will be used again.

Maybe someday,
the calcium will
t r a v e l
halfway
around the world...

...and wind up in the backbone
of a great
blue
whale.